BROKEN M

BROKEN MOON

Carole Satyamurti

Oxford   New York

OXFORD UNIVERSITY PRESS

1987

Oxford University Press, Walton Street, Oxford OX2 6DP

Oxford New York Toronto
Delhi Bombay Calcutta Madras Karachi
Petaling Jaya Singapore Hong Kong Tokyo
Nairobi Dar es Salaam Cape Town
Melbourne Auckland

and associated companies in
Beirut Berlin Ibadan Nicosia

Oxford is a trade mark of Oxford University Press

British Library Cataloguing in Publication Data
Satyamurti, Carole
Broken moon.—(Oxford poets),
I. Title
821'.914 PR6069.A5/
ISBN 0-19-282097-4

Library of Congress Cataloging in Publication Data
Satyamurti, Carole
Broken moon
(Oxford poets)
I. Title. II. Series.
PR6069.A776B7 1987 821'.914 87-12360
ISBN 0-19-282097-4

Set by Rowland Phototypesetting Ltd.
Printed in Great Britain by
J. W. Arrowsmith Ltd., Bristol

To John
without whom each of these would be a poorer thing

ACKNOWLEDGEMENTS

ACKNOWLEDGEMENTS are due to the editors of the following magazines, in which some of these poems first appeared: *Encounter, Grand Piano, Iron, The Listener, London Magazine, Other Poetry, Pennine Platform, PN Review, Poetry Review, Prospice, The Rialto, Times Literary Supplement, Weyfarers, Women's Review, Writing Women*.

The first poem in the sequence 'Between the Lines' won first prize in the 1986 National Poetry Competition.

CONTENTS

Between the Lines	1
Erdywurble	6
Mouthfuls	7
My First Cup of Coffee	8
Clockwork	9
Curtains	10
Letter from Home	11
Pictograph in Dust	12
Coastguard's Son	14
Broken Moon	15
Prognoses	16
Intensive Care	17
Getting There	18
Old Times	20
Mother's Girl	21
Singapore 1963	22
Plums and Porcelain	23
Family Planning	24
Girls' Talk	25
The Uncertainty of the Poet	26
The Archbishop and the Cardinal	28
Feast of Corpus Christi: Warsaw	30
James Harrington's Lament	31
Touch Stone	32
Vertigo	33
Miracle in the Sea of Galilee	34
Letter from Szechuan	35
Balancing Accounts	36
Poppies	37
Women Walking	38
Day Trip	39
Going up the Line: Flanders	40
War Photographer	41
Graffiti	42
From Rosa in São Martinho	43

BETWEEN THE LINES

Words were dust-sheets, blinds.
People dying randomly, for 'want of breath',
shadowed my bed-times.
Babies happened;
adults buried questions under bushes.

Nouns would have been too robust
for body-parts; they were
curt, homeless prepositions—'inside',
'down there', 'behind', 'below'. No word
for what went on in darkness, overheard.

Underground, straining for language
that would let me out, I pressed to the radio,
read forbidden books. And once
visited Mr Cole. His seventeen
budgerigars praised God continually.

He loved all words, he said, though he used
few to force a kiss. All that summer
I longed to ask my mother, starved myself,
prayed, imagined skirts were getting tight,
hoped jumping down ten stairs would put it right.

My parents fought in other rooms,
their tight-lipped murmuring muffled
by flock wallpaper.
What was wrong, what they had to say
couldn't be shared with me.

He crossed the threshold in a wordless
slam of doors. 'Gone to live near work'
my mother said, before she tracked down
my diary, broke the lock, made me cut out
pages that guessed what silence was about.

Summer, light at five. I wake, cold,
steal up the attic stairs,
ease myself into Mrs Dowden's bed,
her mumble settling to snores again.

In a tumbler, teeth enlarged by water;
her profile worrying, a shrunken mask.
Her body's warm, though,
smells of soap and raisins.

I burrow in her arm's deep flesh
forgetting, comforted. Finding out
with fingers that creep like stains
that nipples can be hard as pencil ends,

breasts spongy, vaster than a hand's span;
and further down under the nightdress,
a coarseness, an absence;
not what I'd imagined.

3

Chum Larner, Old Contemptible,
badge on his lapel, barbered our hedges
for parade; nipped capers off nasturtiums,
their peppery juice evoking India,
the dysentery that wiped out his platoon,
'But yer can't kill orf a Cockney sparrer!'
His epic stories rattled gunshot,
showed me what dying meant.

The shed—tropical dusk, air thickened
by tarred twine, drying rosemary,
his onion sandwiches. Sitting on his knee
I'd shiver as he told about the ghost
of Major Armstrong's fancy woman
who wandered the cantonment, crooning.
My mother stopped me going,
suspecting him, perhaps, of more than stories.

4

Upstairs was church,
a clock ticking somewhere
and my mother, a penitent,
breasting the stairs,
smile upside-down,
streak of tears
gone when I looked again.

No sound behind the door
I dawdled past on tip-toe,
where strangers were allowed
and where, wanting more than breath,
my grandmother was being dulled
by blue walls, too much sleep,
the brownish, disinfected smell;
by being too delicate to touch,
by no one singing to her,
by hunger, chafing her to bone.

5

' . . . someone for you to play with.'
But I could tell she'd be
useless at throwing, or bricks,
no good at pretending.

'Isn't she sweet?' Couldn't they see
she was yellow, creased, spotty,
an unfinished frog, a leaky
croaky cry-for-nothing?

'Be gentle now!' But I was
doing what they said, playing.
My best doll's bonnet
fitted her floppy beetroot head.

She smelled of powdered egg.
They warmed her vests.
She slept against my mother's skin.
'How do you like her?' Send her back.

6

Books let me move close to them,
breathe their scent of secret places.
They broke through net curtains,
stained glass, tea-time
and dared to shout, take risks
for love or principle.
One was an unsuitable companion,
revealed with disappointing diagrams
the body's *terra incognita*.

People were difficult to read,
foreign, with uncut pages.
I could see their spines braced
to support the weight of hidden words.
Their covers carried all the information
they thought necessary.

In my dream, I always woke
just as I reached to touch
the most beautiful, the only book
that would have shown me everything.

7

Did I know what I'd find?
I see myself
sneaking, because I must,
across her rose carpet;
turned cat, criminal
without the nerve.

I'm sliding open
the middle drawer
—the letters, wrong
in this habitat of hair-grips,
powder leaves,
the smell of evening gloves.

I can still see
his love-shaped writing,
ink confidently black,
vowels generous
as mouths, fluent
underlined endearments.

And my father's
pinched hieroglyphics,
clipped sentences
the shape of pain
trying to be dignified.
Words pulled out by the root.

Has hindsight twisted it?
I remember no surprise,
only the lurch of knowing
here was the edge
of something absurd,
a terminal complexity.

ERDYWURBLE

My father's parents sold fish.
At school, Greek scholars taunted him,
the scholarship boy,
called him 'bromos', said he stank of fish.
His gifts withered; he learned
a stammer that stayed with him for life,
words jumping like the tiddlers he tried to catch
in the canal.

But from the fractured syllables, there grew
words of his own: 'Don't arrap',
he'd say when we were plaguing him.
'Pass me the erdywurble'—we in giggles
guessing what it was. 'I'm mogadored'
when the last crossword clue eluded him.
'It won't ackle', trying to splint
a broken geranium.

Unable to persuade the doctor
to help him die while he still knew himself,
his words trickled, stopped. Keening continually,
he stumbled on, mistaking night for day,
my mother for his own,
then recognizing no one. Just once,
answering his new granddaughter's cry, he said
'poor kippet'.

MOUTHFULS

They lasted longer then.
Mars Bar paper crackled
as we rewrapped half for later,
sliced the rest
to thin cross-sections,
arranged them like wedding-cake
—loaves and fishes.

Sherbet lemons, hard against the palate,
vicious yellow. Strong sucking
made them spurt, fizz, foam,
sugar splinters lacerate
the inside of my cheeks,
surprising as ice crystals in the wind
that cut my legs through socks.

Licorice comfits shaken in a tin
made marching music.
Or they were fairy food
—each colour wrought a different magic:
mauve for shrinking,
green, the power to fly,
red, the brightest, eternal sleep.

The oddity of gob-stoppers:
tonguing each detail
of the surface—porcelain,
tiny roughnesses,
licking, rolling it, recapturing
the grain and silk of nipple;
rainbows glimpsed only in mirrors.

A shorter life for jelly babies
—drafted into armies, black ones last,
or wrapped in paper shawls in matchbox beds,
taken out, chewed from the feet up,
decapitated out of kindness
or, squeamishly sucked,
reduced to embryos.

MY FIRST CUP OF COFFEE

I'm sophisticated in my Cuban heels,
my mother's blue felt hat
with the smart feather like a fishing fly

as I sit with her in the Kardomah; and
coffee please, I say, not orange squash,
crossing my legs, elegant as an advert.

Beyond the ridges of my mother's perm
the High Street is a silent film
bustling with extras: hands grasping purses,

steering prams, eyes fixed on lists,
bolster hips in safe-choice-coloured skirts
—and then, centre screen, Nicolette Hawkins

(best in the class at hockey, worst at French)
and a boy—kissing,
blouse straining, hands

where they shouldn't be:
the grown-up thing. My hat's hot, silly;
coffee tastes like rust.

My mother, following my gaze, frowns: common.
I'm thinking, if I could do all that
I could be bad at French.

CLOCKWORK

For years, he walked the dog,
his body an espalier,
one shoulder strained back
the other dropped,
dragged onward, always,
leash looped around his wrist,
chain punishing his palm.

His boots rang on the pavements of our town.
They marked the seasons—resonance in autumn
gentled by leaves. The miles he walked
enough to trace the Ganges to its source,
though the dog's route,
like tangled string,
never went beyond the railway line.

No scope, no time for choice,
for planning the journey, asking why;
the dog was responsible.
It was as if
he had been made a toy,
set upon earth to run
harnessed to this inexhaustible clockwork.

And yet, the dog did not exist.
He ran the streets alone,
hand clenched upon itself, responding
to imaginary force. 'Our local eccentric.'
We'd smile as we walked briskly
for our trains; though his hound
loped through our dreams at night.

CURTAINS

Crocheted
they censored light,
grudging flowers
stippled on the wall.

Warped hems
sealed in silence,
empty formalities
of clocks.

Steeped in vinegar,
bleached too decent
for pity or contempt,
they were veils

knotted
against curiosity,
worn enemies
of easy come and go

though cold forced entry,
its fingers
tarnishing every surface
of the room.

Perhaps they were
nourished
by the salt vapours
of her misery:

after she'd gone
their patterns
crumbled
in our hands.

LETTER FROM HOME

Dear Krishnan,

The Chinese should have their own gardens of rest
where they wouldn't disturb decent ceremonies.

Yesterday, Granny Vaithilingam was cremated.
It was very nice—she would have enjoyed it—
until, as we waited for her ashes
in the orchid garden (so pretty
it quite distracts you from the chimneys),
looking to see if Mrs Govindaswamy
had bothered to send a floral tribute,

there was a sudden cacophony of brass instruments
—you couldn't call it music—
and a Chinese funeral party arrived;
such garish clothes,
waving large photographs of the deceased,
chattering, banging gongs. No respect
—one of them trod on Uncle Gopal's wreath.

When the band struck up
'Happy Days Are Here Again'
Auntie and I thought it was time to go.
We were given Granny in a pretty pink box
and, as we left, we heard them playing
'Smoke Gets in Your Eyes'.
Auntie was upset.

We scattered Granny's ashes on Changi beach.
At the last, wandering,
she had said she would prefer
to be sprinkled on Serangoon Road
outside Komala Vilas, where they make
such delicious dosais.
But we didn't feel it would be very nice.

Please take care, don't neglect your studies;
wrap up well against the London fog.

> With fondest love,
> Mother.

PICTOGRAPH IN DUST

Our land has forgotten the taste of rain,
the sky hot, scorning us for years.
We wander, settle for a time,
build houses round ourselves,
cut doors out last.

White men came on roaring carts,
showed us by signs
a different kind of place
where water leaps out of the earth
and we could live soft always.

But this is where we grew.
We are dry people, deep-rooted as thorns,
baked like our cooking-pots.
The earth holds the shape of our heels;
our ancestors need our songs.

They pointed at the sky,
played frightened, waved their arms,
then shook their heads, went away.
The land threw dust
into the air behind them.

Three dawns. Sky flash. An extra sun,
a monstrous cloud, beautiful as rain-dreams,
blossoming. We lost ourselves in looking,
lost our skin, our hair.
Was this what they were pointing to?

And lately, a new sickness.
The strangeness of it made us weep
until the elders spoke:
'All death is one,
only the tracks we take to it are different.'

Could we scratch pictures,
tell people who come after us, and after,
how the white men's spirits are terrible
to those who raise their eyes
above the thorns?

We are building our last houses
—as we have always built
but with no doors. We shall grow light,
crumble like earthenware,
become the land.

COASTGUARD'S SON

(after Webster Smalley)

When the dog died, slowly,
entrails glistening in the sun,
eyes fixed on his until they filmed,
the boy cried gritty tears,
swore he'd never love a creature
he couldn't save from pain.

But when he saw the whale
leap in slow arcs, dwarfing the pines,
rainbows in the torrents
crashing from her back,
he loved her, though he knew
moloch whalers prowled beyond the strait.

He plunged his head into the sea,
heard her voice, unanswered;
surfaced, gasping,
found an old pipe to listen through.
He lay for nights,
forehead strained against the sand

until the notes made patterns,
he grasped the hidden grammar of her song
and, mouth pressed to the pipe,
he called out, in her tongue,
his fears and curiosity and love;
and heard her answer him.

BROKEN MOON
for Emma

Twelve, small as six,
strength, movement, hearing
all given in half measure,
my daughter,
child of genetic carelessness,
walks uphill, always.

I watch her morning face;
precocious patience as she hooks each sock,
creeps it up her foot,
aims her jersey like a quoit.
My fingers twitch;
her private frown deters.

Her jokes can sting:
'My life is like dressed crab
—lot of effort, rather little meat.'
Yet she delights in seedlings taking root,
finding a fossil,
a surprise dessert.

Chopin will not yield to her stiff touch;
I hear her cursing.
She paces Bach exactly,
firm rounding of perfect cadences.
Somewhere inside
she is dancing a courante.

In dreams she skims the sand,
curls toes into the ooze of pools,
leaps on to stanchions.
Awake, her cousins take her hands;
they lean into the waves,
stick-child between curved sturdiness.

She turns away from stares,
laughs at the boy who asks
if she will find a midget husband.
Ten years ago, cradling her,
I showed her the slice of silver in the sky.
'Moon broken', she said.

'She'll walk something like this . . .'
Springing from his chair
he waddles, knees crumpled,
on the outer edges of his feet
—a hunchback, jester, ape,
a wind-up toy
assembled by a saboteur.
I turn away, concentrate
on the caesarean sting.

I wander corridors.

Far off, approaching,
a couple, hand in hand,
the girl, lurching
against the window's light.
I hear them laugh, pick up
the drift—a private joke,
the film they saw last night.
Long after they are gone, I hear
the jaunty click-creak of her calipers.

INTENSIVE CARE

Your voice silenced by tubes,
the mute, continual cough lifts you awake.
I stroke your hair; you stare at me,
eyes remote, tearless.

You write, 'I'm hungry'.
I watch each breath
sucked in between your ribs,
beg for you.

You lie as if in state,
too dignified.
If I thought you were leaving me
from this white room

with only plastic pillows for your journey
I would cram your hands with anemones,
snatch out the catheters, enfold you,
run with you to where the band is playing.

But now, as my hands
make shadow creatures on the wall,
I read your lips: 'rhinoceros',
know I have you still.

GETTING THERE

Sports Day. Miss Cook
had whispered to the rest
to let you win the walking race
and not to tell.

Your jerky gait,
your straining;
the others shuffling behind,
their over-hearty cheers,

you pleased, unsure.
When your friend confessed,
wanting you to be
like anyone again,

you looked bereft,
confused
as if the walls
had changed alignment.

You understand proportion.
I still wake at night
explaining to Miss Cook
why she was wrong

—I know the artfulness
of happy endings:
once when you were small,
still chair-bound,

I dreamed you walked
perfectly into my room;
somehow, even in the dream,
a counterfeit—but so real

I woke shaking, as though
I'd almost been drawn
into a lotus-land
where I'd never find you.

Sometimes, when we're gay,
we hold hands, polka round
like dancing bears,
laughing at each other.

At the table, in heat-heavy shade
we smile away absence, drink wine
'like Italy . . .' and you recall your honeymoon
at Garda, fascisti marching,
tumbling geraniums, lake's unexpected chill.
Abstracted, you over-fill your glass,
meniscus trembling: tears at eyelid's brink.

We carry lunch outside through well-used rooms,
I following your plod round hazards
of rucked rugs. Her photograph more formal
since she died, she stares through dust
at tipsy piles of books, turmoil of print.
The whole house has become your library;
has she grown to find the clutter comforting?

We feast on meat, fresh bread, neglected lettuce
from your garden; flavours enhanced by friendship
—even by absently swatted flies you peel
from mortadella. You say that soon
the time will come when you must burn papers
you'd not want anyone to read. Letters?
Diaries, in which you answered back?

'The house seems oddly silent, even now.'
But five years' dust has partly blanketed
the pain, furred memory's uncomfortable edge;
you have been making bookshelves since the spring.
As I leave, smoke from your panatella
weaves wraiths in the warm dusk. You kiss me,
ask, 'and do you play her viola still?'

MOTHER'S GIRL

(i.m. *Pat Bain*)

She remembers a mother waving from a train;
'Don't cry, silly girl,
Mother will come back very soon.'

As her life leaches into the still air
she watches tramps shuffle from the park,
knows envy's vertigo.

She hears herself speak clichés: ' . . . a nightmare',
sees friends look reassured that dying
can be compared to anything at all familiar.

'The children will remember me ugly.'
But she is bone-beautiful, a Giacometti,
filigree of veins in yellowed ivory.

She wears parrot colours;
she buys great bags of tulip bulbs,
learns a new Berlioz song, talks of a holiday.

While her husband whispers to the children
she turns the fragile vaulting of her back;
marzipan smile crumbles, tastes of quinine.

A stranger's fingers clutch the furniture
—splinters, fat enough last month to draw
bravura from the piano, coax a baby into sleep.

Dressed for the ward-round, she is actress
and audience, hair in a bright bandana,
watching through ice, marionettes, miming.

Roused by a small hand from morphine dreaming
she murmurs, as she sails the summer night,
'Mother will be better, very soon.'

'The overseas Chinese' you said,
rehearsing your lecture as we walked,
'lack revolutionary potential
—ticks on the hide of the bourgeoisie.'

People who seemed androgynous with age
haunted the docks, the streets,
shuffled like exhausted ballet dancers
from shadows, to the lee of ships, to archways,
waiting to scize what nobody would miss.

'Even the beggars are capitalists at heart.'

In the naphtha glare of the night market
a woman spread a mat, laid out her stock:
two apples, a plastic pack of nuts,
a reel of sewing thread,
a dog-eared birthday card with daffodils,
assorted nails.
We wouldn't be her customers;
her cat-cries aimed at those
almost as ragged as herself.

'The problem is their thorough individualism.'

A history of sharing shallow bowls,
a pipe. In good times,
a few hours' bed space
in Sago Street.

We lost contact. I heard about your chair
in another newly independent country.
Your monograph on African Socialism
had excellent reviews.

PLUMS AND PORCELAIN

Plums glisten
on porcelain plate.
Mouth sized. Skin taut

as shoulders—labourers
crowded close,
pearl black, gleaming

curves of shoulders
where the arm begins,
shaped for cupped hands.

Plum-in-mouth, Master
barks orders, claps hands,
shoulders seamed, angular;

stiff shorts, scuffed knees
poignant as a schoolboy's
against his rifle's sheen.

Memsahib yawns in shade,
shoulders puffed muslin,
hand cupped round a plum.

FAMILY PLANNING

Here is my clutch of humbugs, fickle honey-bees
swarming, sedate. Pleased to see me
—aren't you, my fondant fancies?
I want you filling every hollow of my house;
I need more. More.
Another journey to the suburbs.

I wait till after midnight,
watch bedroom lights shut off, slip
down side passages, over well-clipped lawns
searching for them. Plucking them from walls
and window sills, I tickle their ears
with tales of fish heads, drop them in my basket,
close the lid. You welcome strangers coolly
my wasps, my soft moss-agates.

I have founded a dynasty. If these pharaohs
quarrel, they turn into racoons,
tails fat with malice; I croon them
into sulky tolerance. As it grows dark
they seethe across the flagstones;
a hundred phosphorescent pools
spangle the night courtyard, pitiless.
Hunt well, my predatory loves.

At dawn the kitchen air is heavy,
moist with cat breath; on the range
the mound heaves gently,
until the hot ones struggle from beneath,
clamber on top—as, had I married,
my children might have played hand sandwiches.

GIRLS' TALK

Then Miss Rodway sent the boys
out to the playground, and she told us
about the German measles injection
so we won't have deformed babies.

She said we could ask her
anything we liked. She twisted the ring
on her finger round and round,
swinging her legs.

I asked if there's a wrong way
to have sex; if the sperms always
find the way, or if they sometimes
get lost, spill on to the sheets.

The boys were jumping up outside the window;
Miss made cross shooing faces.
She said men have stronger urges,
that's why many marriages don't last.

Natalie's got periods already
even though she plays football.
Mr Davis won't let her go to the toilet
when she asks, because she's naughty.

She told us about the pill
but I wasn't listening—I was thinking
of what I do in bed at night
when bad dreams come out of the cupboard.

It wasn't the same as when you told me
—it seemed so serious
Catherine and I couldn't stop laughing;
we stuffed tissues in our mouths.

But Mum, I'm scared of the injection.

THE UNCERTAINTY OF THE POET

(after de Chirico)

1

Is there no answer to sexual obsession
—humiliation of this clown that won't lie down
but leads me to jump absurd
into unsuitable beds, leaving the Muse unserved?
Oh, cul-de-sac delusions, love sickness
that warps imagination, subverts art.

The Corybantes knew the way of it,
those self-made eunuchs in Cybele's name
—well, my neglected and neglectful Muse,
I'll go to Dayton, Ohio. There they trap
the lecher in the brain, lobotomize . . .
a certain way to end the sabotage.

2

The organ withers, sleeps. My work
will flower—no women squeezing juices
that should be the Muse's. But these breasts
and broadened hips I've grown are disconcerting,
and it hurts when, hugging old friends,
I see myself reflected in their frozen eyes.

Arms, hands are a distraction. I've never doubted
that losing all libido for Her sake
will liberate great verses from my pen.
But when? My fingers ache to stroke warm flesh,
plant trees, shape earthen figures. The first step
made others easier; I'll ring Dayton, Ohio.

3

Without arms, I no longer had to play at respectable
employment. I could embrace (so to speak) the role of poet.
But bureaucrats refused to pension me
while I could walk. I could be a traffic warden,
'pleasant outdoor work'. I'd write tickets with my teeth;
they'd give a tactfully remodelled uniform.

So I've had my legs removed. It was time in any case
—for months I'd been unable to sit still. Each time
I settled at my table—torment; an unappeasable desire
to dance, to walk a tightrope, clamber up rocks,
dabble in foot painting . . . I'm one of their most
rewarding clients they tell me, here in Dayton, Ohio.

4

It's winter. They've given me a room looking out
over the plains, where I can write uninterrupted
on my remote-control word processor. Ideas spin
prosy patterns, images inane as ticker-tape,
streams of dead metaphors. Why is there no spark?
If I asked them to excise the intellect,

all senses, leaving only the heart, could I achieve
sensibility distilled, the perfect poem? Worth the price
of not being able to impart it—except to Her.
The decapitation fee here at Dayton, Ohio,
is extremely high. But I can sell my house,
my books, my—everything. I shan't be needing them.

5

They have positioned me on the terrace. I can feel sun
on my skin, though I am cold. I sense the vibration of
a distant train: the Muse, perhaps, leaving for a more
fertile climate? As my pulse slows, syncopated, I imagine, next to me, phantasms of poems, clustered, like
fruits: gold, growing vigorously from the central stem.
One, broken off

Two old men
stand in the palace garden
in their dressing-up clothes.

Hats architectural
—one a dome, the other
flying buttressed—garments

velvet, watered silk,
loops of gold chain,
these walking oxymorons

wear curious expressions
—as though the Cardinal's
found something more delicious

than the fifth deadly sin
to which his ample chasuble
bears witness

(greater ecumenism hath no man
than to share
a certain kind of secret).

In Lambeth back-streets
rack-rented tenants
spill on to landings.

In Liverpool
too many children
drive women mad,

while here, two old men
amble on the lawn,
landlord, father by proxy

—though it's not personal.
He might have been
a country schoolmaster

and he, with his
ruddy, potato face,
a labourer.

And these starched linens,
fine worsteds, could stand
empty, gossiping together.

The line snail-ribbons
down Krakowskie Street—women,
girls, some medalled veterans.
Slow hymns, as for a funeral;
crowds press, voices join in,
a helicopter tacking overhead.

The procession swerves,
passes the crucifix of flowers
made secretly one night, by women:
flowers as witness,
candles for endurance,
lumps of coal for solidarity.

Candles in jars, steady in the draught;
soldiers with Modigliani faces.
Is it their grandmothers
who bring fresh flowers each day,
work them in calmly, eyes lowered
as their knees roughen on stone?

It could be their sisters,
surge of first-communion white,
who know the hymns by heart,
who bear these banners
embroidered with an image from the East:
a Madonna, black as coal.

JAMES HARRINGTON'S LAMENT

*(James Harrington, the seventeenth-century political philoso-
pher, when imprisoned in the Tower, believed that his sweat was
turning into bees)*

New life—out of the swamp of flesh,
emerging damp as swimmers
on a surge of tide, cast up, inert
until a tremor, uncertain singing,
a frither of wings confirms the miracle.

They swell, fill out as apples do
from ruins of limp flowers,
then show their separate power, detached
as apples from the tree—a fellowship
of strangers, no longer needing me.

Oh, this fruition would make me glad
but for the whisper of ambition
that this be just the start, that I'll go on
to vomit doves, weep fish,
expel children from my deepest orifice.

For knowing the leap of one becoming two
we punish women. And, not to face our grief,
we chase cold secrets of the galaxy,
plumb oceans, raise spires, shivering at dreams
that make of our greatest doings, clumsy tricks.

The names of your stone—agate,
chalcedony, quartz—are spell-words.
Touching it, blind, might yield
the cool brush of your hand.

Contoured tracks of polished ochre,
sap-white, grey, frame
a heart-shaped dance of splinters,
opaque until, held to light,

it owns to flaws,
frozen secrets
glimpses of fish, of embryos,
deep patterns in your eyes.

Gelatinous, it seems
that it could melt, assume new forms
yet remain itself;
marbling on a stretched balloon.

I trace its rough edge
thinking how stone
we once touched in each other
turns to amber.

VERTIGO

If I should start to think too vividly
of how, while I lie here, tossing for rest,
enduring night, you, earthed at a different angle,
sit on a pine-clad hill, miles to the west

and paint the sun in wine, gaze out at peaks
adorned by ancient names; or, drunk with talk
of old times with old friends becoming old,
circle your finger with a whiskered stalk

of Rocky Mountain poppy, I should lose
my balance, slide to childhood make-believe,
step off the world's edge, plummet, fly apart
and carried, senseless, in the wind's wide sleeve,

atoms of me might fall in foreign rain
defying odds, in touch with you again.

In pampered pilgrimage
we drive, following the map,
to Galilee. Town-soft,
our feet are bruised by stones
until, deep water lifting us,
we strike out, separate.

The water chills under its skin
so we are vigorous.
Flight of your white arm
extending on the edge of vision.
I turn to you, flip away;
Mount Tabor seems less distant.

Returning to shallows I lie still,
nuzzled by fish,
sun comforting my shoulders;
fish babies, mother-tied invisibly,
risk jerky, brief adventures
among flecked stones.

I collect *famille rose*
—spectrum of flesh pigments,
shapes sugared-almond smooth.
Pinks fine-veined as petals,
pale orange glisten of roe, terracotta,
burnt umber: colours born in water.

Suppose we should turn into pebbles
coloured like earth
baked hard, fire-simplified,
and lose all angularities
in the ebb and flood,
pain-free, indifferent . . .

My fingers close on something conical.
I look up, find you near;
we open our hands
in yours, as in mine,
out of a million stones,
a tiny, living shell.

LETTER FROM SZECHUAN

You cannot prevent the birds of sorrow flying over your head, but you can prevent them building nests in your hair.
(Chinese proverb)

Wisdom of fools and schoolmasters
—men whose heads are cabinets of drawers,
contents wax-sealed.
No purchase there for birds;
claws, sliding over perfect lacquer,
rattle the handles, bring, for a moment only,
a flutter to the dark interiors.

But we are made differently, Yi Lin.
If I could tell you of the birds,
how they have settled with me
since you were taken,
I know you would say, 'And I, and I!'
Let me write as if . . .

That first morning, having slept at last,
I woke to a jostle of high notes.
I felt their weight, talons
grasping loops of hair,
twisting for footholds, already tangling
strands into a nest;
crying—a narrow song,
a canon without resolution.

I have learned their ways.
Though I cannot see them, I know
they are the colour of tarnished laughter;
eggs heavy, leather-surfaced, rough.
They are dim-witted rather than malevolent
—pecking through my skull, not seeing
that when they have devoured hope, memory,
they will be homeless.

Birds of happiness have many songs,
these only one—my friend,
in whatever province you are lying
they sing it for you too.

She's packed
ready to lose him
at a moment's notice.
Marching orders come
in that slight stammer
she's loved so much
in words she knows already.

She starts to speak.
He glances at the clock,
a habit she's trimmed to.
She draws together
all traces of herself.
She has a train to catch.

Or else

the sense of that bag
waiting, seeking attention,
its silent provocation
like someone
turning the first cheek,
weighs with him increasingly.

He knows her need
for ends tied up,
her inability to wear guilt
gracefully. Generous,
he sends her packing;
only, he'd be undone
by talk.

POPPIES

He used arrive without no warnin'
just phone from somewhere
on the motorway. Hurry, quick
put on fresh sheets
run to Patel's—sausages, white bread
chocolate biscuits
(I think his wife a healthy livin' lady)
grab poppies from the yard
stuff them in a glass
put on dress he say he like once.

Sit and shiver. Afraid I ugly,
afraid his face fall, look aside;
no words—he don't want me
chattin' on, with him a swallow
swoopin' all over on the motorways.
Each time I forget he talk so easy.

Stories! People I never see
dance colours on the empty wall;
he make me laugh like never,
he make the stories loosen in me
only I too shy. It get late
and now poppies droopin', but he not.
He really like me, it me he lookin' at
like it the first time always.
He stroke my face, breasts, like wonder,
soft kiss my lips so they perfect.

That last time he say he love me.
He surprise as me—we both laugh.
He not come again. Ever since,
I dreamin' often I lyin' in the yard
can't move nothin', and my nipples
blossomin' with poppies.

My day is fettered by my mother's steps.
I learn the shopping list by heart,
discover architraves.
Walking this slowly
I nearly lose my balance.
I've not got that long—
at my pace I'd be going
somewhere, not marking time,
her arm locked on to mine.

*

My daughter's somewhere else.
Her tenseness fusses me
into unsteadiness.
Her arm is wooden.
Once there was suppleness,
a give and take,
a comfortable distance.
I didn't ask for this—
time, pace, speed, out of my hands.

*

Haven't we walked this way before
—a child fumbling, breathless,
clutching to keep up;
a mother tethered to a clinging hand?

DAY TRIP

Two women, seventies, hold hands
on the edge of Essex,
hair in strong nets,
shrieked laughter echoing gulls
as shingle sucks from under feet
easing in brine.

There must be an unspoken point
when the sea feels like
their future. No longer paddling,
ankles submerge in lace,
in satin ripple.
Dress hems darken.

They do not risk their balance
for the shimmering of ships
at the horizon's sweep
as, thigh deep, they inch on
fingers splayed, wrists bent,
learning to walk again.

Mme Verklaede, mother of four tall sons,
hangs out washing on a fine drying day,
shirt after shirt facing the same way,
off on their anchored dance.

One, swollen with bravado,
advances towards the sky;
another writhes, reluctant to yield
to the sun's shifty blandishments.
This tattered one, a plaything for the cat,
draggles its limp sleeve along the grass.
While that one hangs crucified,
its striped brother, made of different stuff,
clowns in frantic acrobatics.
Another catches its hem on rose-thorns,
resists the summons of the wind
that makes its neighbours chatter.

Here, from beneath our feet
—were there an instrument patient enough
to tease messages along the threads—
we could exhume the uniforms,
scrape off mud, tip out the bones,
reconstitute the men who hung on them.
The biography of one nineteen-year-old
would stretch for miles
telling how he shivered that July,
played cards, wrote half-truths home,
clutched a frail talisman inside his tunic,
faint with heart-beats louder than the shells.

Mme Verklaede starts to gather up
and fold her wind-threshed harvest.
A calm evening; a faint breeze from the west
carries the bugle: the last post, from Ypres.

The reassurance of the frame is flexible
—you can think that just outside it
people eat, sleep, love normally
while I seek out the tragic, the absurd,
to make a subject.
Or if the picture's such as lifts the heart
the firmness of the edges can convince you
this is how things are

—as when at Ascot once
I took a pair of peach, sun-gilded girls
rolling, silk-crumpled, on the grass
in champagne giggles

—as last week, when I followed a small girl
staggering down some devastated street,
hip thrust out under a baby's weight.
She saw me seeing her; my finger pressed.

At the corner, the first bomb of the morning
shattered the stones.
Instinct prevailing, she dropped her burden
and, mouth too small for her dark scream,
began to run . . .

The picture showed the little mother
the almost-smile. Their caption read
'Even in hell the human spirit
triumphs over all.'
But hell, like heaven, is untidy,
its boundaries
arbitrary as a blood stain on a wall.

Paper having acquired a poor image,
they each in turn took a diamond stylus,
signed the treaty on a sheet of glass,
words clear against a background of red or blue.

They posed for the cameras, hair lacquered,
identical suits, each middle button fastened
to conceal the rate of respiration
and prevent any unplanned flapping of the tie.

They shook hands in ritually prescribed order,
crossing fingers in left trouser pockets
to neutralize untruth. Smiles locked in place
they saw the blood in one another's eyes.

They put on their spectacles, made speeches
(mutually incomprehensible, all equally sincere)
broadcast to the world
a sense of their historic destiny.

Then they flew home, unbuttoning their suits.
One had inscribed his name in mirror-writing.
Later, when the treaty was overturned,
he was found to be the only one on the right side.

FROM ROSA IN SÃO MARTINHO

for Maria Pinto

1 *Postcard*

Looping the coast
—mountains, glint of levadas,
banana groves. So many houses!
Touch-down,
finding my face wet.

Shrieks, embraces, presents,
peeping neighbours,
maracujà, honey cake
—and noise! Night:
a lizard winking.

It has to be like this—
feeling my way
through grittiness of soap,
enamel plates, back
into the textures of home.

Not the place I fled from—
this is a peeled, harmless replica.
I'm back, but almost as a tourist;
I don't need, after all,
the city clothes, new black suitcase
to ward off the past.

It's not that I've forgotten my father:
the burn of leather belt on skin,
fear drying the mouth like quince.
It's not that the colours—wine flush
of his eyes, neck's purple veins—
have blurred at all.

I remember how I spat him out, turned
a scarred back against him. Now, I know
I simplified him, censored the vision
of his head—pale strip of forehead
bared to the landlord, hat awkward
in his dirt-stiff hands.

But the memories are flat,
scissored frames from a lost film.
I left in silence, refused
to ask his blessing. Today, easily,
bending to kiss this tearful stranger,
I whisper, 'Pai, sua benção.'

3 *Embroidery*

All day I sit cross-legged with the women
embroidering, talking of husbands
—past, present and to be arranged.
Sisters, cousins, aunts, we make the flowers
our mothers showed us, white on white.

Curved spines, rough ankles,
flattened finger-ends—their bodies
moulded to the task, they pull their threads
taut, shape disappointments
into an appliqué of laughter.

At night I hold a phantom needle,
feel my arm still lifting thread, falling.
My eye sees templates everywhere
—the sea marked out with lights of tiny boats,
the sky pricked by stars.

4 Patching

No real men here—
only the hopeless stay,
those softened young
by wine for wages—empty men
grown thin as clothes pegs.

Around six yesterday, my sister
laid down her needlework,
took it up, her stitches crazy,
shouted at the children,
stood waiting by the road

for Alfredo, late, staggering
arrogant as a toddler, bawling
a vicious song, wanting her
to punish. She, twice his weight,
allowed him to be strong.

Today, he stayed at home. She hid
her bruises, unpicked stitches,
fed him baby-soup. Tonight, he sits
silent, smoking harsh tobacco,
turning five escudos in his hand.

5 *Futures*

My niece walks with me in moonlight:
'I'll marry an Italian, like my aunt.
He'll be tall, blonde perhaps.
We'll ride a gondola. I'll have
a silver kitchen that works by itself.'
I squeeze her hand. I know
her mother has an eye on Paolo,
neighbour's boy—short, quiet,
working in an office in Machico.
And Fernanda's a sensible girl.

Keen for my reaction, my nephew tells me
he likes calculus. He's quick,
drums his fingers, restless for something.
I imagine him a scientist—know
he'll leave school when he can,
work for an uncle who makes coffins.
Already, in the way he turns his head,
his hooded look, angering his sisters,
there's the old pride that draws on nothing
but itself, and ends by drowning.

Shouting, they heave the dead weights
up the ramp, scales flashing,
slap them down on stones, heaped high,
spilling the smell of sea.

Women promenade, size up the catch,
begin the ancient ritual
—clamour for prices, feigned disbelief,
shrewd scrutiny of measures.

The fishermen throw down their caps,
wield hatchets, cleave great tuna
into chunks, rub salt on,
spread them in sun to dry.

My sister shows me off
—her English relative. Neighbours
dissect me with their eyes, whisper
rumours of my past.

Pride of the catch, the black espada,
ugly scabbard fish, leers
as if embarrassed at being caught
dead on a trestle table.

I stare at it: poor oddity.
In my mind's eye, its muscles
leap again; it strikes out, plunges
back to its gypsying.

7 *Photograph*

So that's who she was—
not my collage of gilded fragments,
sugar saint, eyes sea deep,
comforting me, her favourite,
but a plain girl, starved of choices,
whose bones lie hidden somewhere here
anonymous as flints.

In the creased studio photograph
there's pride, a sort of avidness
transfiguring the desperate impatience
my brother says she showed with all of us.
She poses, a star,
embracing her moment
before the shutter snapped.

I have to leave her here,
mother who never was,
be mother to myself.
But I remember reaching up
to hand her clothes-pegs,
laughing with her as we named them
—Manuel, Josè, Vicente, Father João.

This is where we sat
chasing lice
through one another's hair
sucking marrow bones.
Sometimes, after dark,
I'd slip out
to these plum trees,
shrunken now,
gorge stolen fruit.

I thought I'd find the faces,
tarnished veils
that stifled me for years.
But the house is empty,
stripped of the vast
camphor-smelling armoires,
credences, the cornered saints
whose monstrous shadows
subdued our urge to sing.

I hated them for their pinched
insistence on the rules.
Maybe they believed
there was no better language
they could teach a girl
than that of service,
curbing our tongues, hands, eyes
—His will be done. Today
I would have told them otherwise.

9 *Procession*

Christ, in his private ecstasy of pain,
parades the streets. The sign-writer
who retouches the gilt from time to time,
the woman who dusts Him every day
stand breathless as He passes.

Earlier they and many others knelt
as I did once, arranging lilies, agapanthus,
marigolds into a patterned carpet
which now the tubby priest,
the bearers of the statue, trample on.

We have lost touch, He and I.
I can't recapture that straightforward love,
though whether it was time, space
or experience that distanced us, who knows.
My clumsy lips shape hymns, invitations

to a place I can't climb back to.
Against the church wall, not singing,
a blind man stands alone
with outstretched hands
on which rain starts spitting.

The engine throbs;
the island
already
foreign.

The runway
a dark finger
flicking me up,
out to sea.

OXFORD POETS

Fleur Adcock

Yehuda Amichai

James Berry

Edward Kamau Brathwaite

Joseph Brodsky

D. J. Enright

Roy Fisher

David Gascoyne

David Harsent

Anthony Hecht

Zbigniew Herbert

Thomas Kinsella

Brad Leithauser

Herbert Lomas

Derek Mahon

Medbh McGuckian

James Merrill

John Montague

Peter Porter

Craig Raine

Tom Rawling

Christopher Reid

Stephen Romer

Carole Satyamurti

Peter Scupham

Penelope Shuttle

Louis Simpson

Anne Stevenson

Anthony Thwaite

Charles Tomlinson

Andrei Voznesensky

Chris Wallace-Crabbe

Hugo Williams

also

Basil Bunting

Keith Douglas

Edward Thomas